W9-AFD-755

LIFEWATCH

The Mystery of Nature

Seed to Peanut

Oliver S. Owen

Published by Abdo & Daughters, 4940 Viking Drive, Suite 622, Edina, Minnesota 55435.

Printed in the United States.

Cover Photo credit: Grant Heilman Photography, Inc.
Interior Photo credits: Peter Arnold, Inc., Pages 4, 7, 12
Grant Heilman Photography, Inc., pages 5, 8, 10, 14, 17, 19-25

Edited by Bob Italia

Library of Congress cataloging-in-Publication Data

Owen, Oliver S., 1920 Seed to peanut / Oliver S. Owen.
p. cm. — (Lifewatch)
Includes bibliographical references (p. 30) and index.
ISBN 1-56239-489-4
1. Peanuts—Juvenile literature. I. Title. II. Series:
Owen, Oliver S., 1920- Lifewatch.
SB351.P3094 1995
583'.322—dc20 95-10523
CIP
AC

Contents

The Peanut

What would we do without the wonderful taste of peanuts? Maybe you have peanut butter sandwiches for lunch. Peanuts make candy bars special, too. And how about a big bag of salted peanuts at the ball game? As you munch on your peanuts, have you ever wondered what the peanut plant looks like? Where do peanuts grow? How are they harvested? Are peanuts good for something besides food? How can you grow peanuts in your own back yard?

The peanut belongs to the bean family.

The Story of the Peanut

The peanut is not a nut after all! It belongs to the bean and pea family. This explains some of its other names: ground peas and goober peas. In some parts of the United States, peanuts are called goobers, goober nuts and pindas.

Humans discovered the peanut in Argentina more than 5,000 years ago. Peanut farming gradually spread throughout South America. Africa learned about peanuts in the 1500s when the Portuguese brought them over from South America. In the 1600s, peanuts were brought to the United States on slave ships from Africa. President Thomas Jefferson planted peanuts on his Virginia farm in 1781. But at first, peanuts were not popular in America. Early American settlers used peanuts to fatten up their cattle and pigs. In 1917, Americans finally developed a taste for peanuts.

The peanut is not a nut.

The Peanut Plant

The Roots. The peanut plant has a large main root which grows into the soil. It is called a tap root. A number of smaller roots grow sideways from the tap root. These roots help anchor the peanut plant in the soil so that it doesn't blow over in a strong wind. Each root has millions of microscopic hairs. They take in water and nutrients from the soil.

The roots have a large number of roundish swellings called nodules. These nodules are found on peanuts, beans and peas. If you sliced open a nodule and looked at it under the microscope, you would see hundreds of rod-shaped bacteria that use the nodule as their home. These bacteria are valuable to the peanut plant. They change the soil nitrogen so the peanut plant can grow and stay healthy. Some nitrogen passes into the soil and increases its fertility. Some farmers bury the whole plant after the peanuts have been removed so that the soil is enriched. This is called green manuring and saves the farmer fertilizer expense.

The peanut plant bears a large number of branching stems. In the bush type, the branches grow upright. In the runner type, the stems grow close to the ground. The stems have several important jobs. They hold the leaves in the air.

The peanut plant bears a large number of branching stems.

This allows the leaves to exchange gases with the atmosphere. The stems have pipelines which carry food from leaves to hungry parts of the plant, such as the roots and flowers. Other pipelines carry water and nutrients from the roots to the leaves and flowers.

The stems act as pipelines carrying food to the roots and the flowers.

The Leaves. The leaves look much like those of the pea plant. Each leaf is divided into two pairs of leaflets. The leaflets are smooth on top and hairy on the bottom. They are the peanut plant's food factories. To make this food, they use water from the soil. This water travels through the roots and stem and into the leaves.

The leaves also need carbon dioxide gas to make food. They get this gas from the air. It enters the leaf through millions of microscopic breathing pores. The food-making act is powered by the sun's energy. This energy is taken in by millions of tiny green wafer-shaped bodies called chloroplasts ("green bodies"). They contain green pigment called chlorophyll ("green leaf"). The chloroplasts give the leaves their green color.

The chloroplasts give the leaves their green color.

The food-making act in the leaf is called photosynthesis ("put together with light").

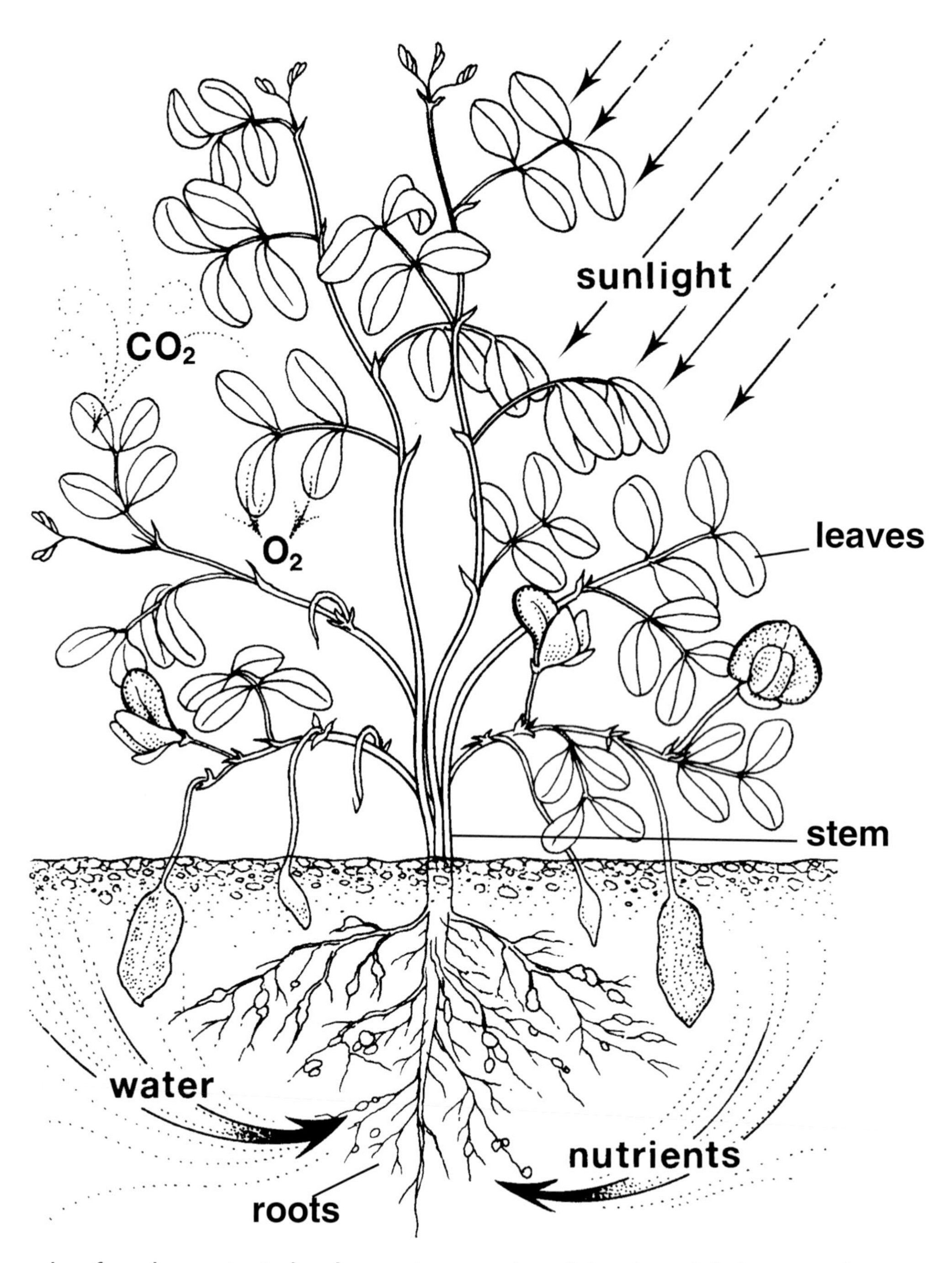

To make food, roots take in water and nutrients which travel through the stem and into the leaves. The leaves use sunlight and carbon dioxide to convert the water and nutrients into food. This is called photosynthesis. The leaves then release oxygen into the air.

During photosynthesis, the carbon dioxide and water is made into energy-rich sugar with the aid of the chlorophyll. The peanut plant burns this sugar to grow and reproduce.

The Flower. The pea plant blooms for about two months. The small flowers have yellow petals with orange stripes. Their life is very short—only a few hours! They open at sunrise, and shrivel and die by mid-day. However, while they are open they do some very important work.

The flowers contain both male and female reproductive organs. The male organs are club-shaped stamens. At the end of each stamen is a sac which makes pollen grains. Each pollen grain contains a male germ cell or sperm. The female organ is located in the center of the flower. It is shaped like a long-necked vase. At the bottom of this vase is the ovary in which the eggs are formed. Honey bees are attracted to the flower's nectar.

As the bees move around they brush pollen from the stamens to the top of the female organ. Then a sperm cell from a pollen grain grows down through the neck of the female organ to the ovary. Once inside the ovary the sperm cell unites with the egg. This is called fertilization. A seed (peanut) will form from each fertilized egg.

The peanut plant blooms for two months. The small flowers have yellow petals.

The stalk which holds the ovary and fertilized eggs is called a peg. The peg grows down to the ground. It then pushes into the soil to a depth of about three inches (7.5 cm). The tip of the peg swells to form a pod or peanut. Each pod is similar to that of a bean or pea plant except that it usually holds only two seeds. A bean or pea pod may have many more.

Peanut Farming

Twenty million tons (nine million metric tons) of peanuts are produced worldwide every year. The leading peanut growing nations are India, China and the United States. Forty thousand U.S. peanut farmers grow more than a million tons (.5 million metric tons) of peanuts yearly. Former President Jimmy Carter owns a large peanut farm near Plains, Georgia. That state alone produces 45 percent of America's peanuts. Other important peanut-producing states are Alabama, fourteen percent; North Carolina, eleven percent; Texas, eleven percent; and Virginia, seven percent.

The plants need about five months of warm, frost-free weather to grow well. They also need 30 inches (75 centimeters) of rainfall yearly. Such weather is found mainly in the Southeastern states.

Before the seeds (peanuts) are planted, the farmer thoroughly loosens the soil. Otherwise the pod cannot force its way through the ground. After five months of growth, the plants are ready for harvest. The plants are plowed up with a sharp-bladed machine and are left in the field to dry in the wind and sun. When thoroughly dry, the farmer harvests the crop with a combine. This machine gathers up the dried plants and collects the peanuts.

Forty thousand U.S. peanut farmers grow more than a million tons (.5 million metric tons) of peanuts yearly.

Southern farmers have found peanut farming profitable. But it was not always that way. In the 1800s, cotton was the major crop. Very few farmers raised peanuts. But the son of a black slave changed all that. His name was George Washington Carver.

Carver was born in Diamond Grove, Missouri. While in grade school, he became interested in plants. His classmates thought he was a little strange. But they respected him. They called him "the plant doctor." As he got older, his interest in plants grew stronger. When he was 37 years old, Carver was invited to do research at Alabama's famous Tuskegee Institute. There he tried to find other uses for peanuts besides food. He felt that southern farmers should plant more peanuts. Then they would not be wiped out if insects or disease destroyed their cotton crop.

Carver found at least 300 uses for peanuts, including the "coffee" he drank for breakfast. Farmers got excited about this lowly nut. More and more peanut acres were planted. After a few years, peanuts became one of the South's biggest crops.

Use of Peanuts

Food Value. The raw untreated peanut fresh off the plant is not very tasty. More than 25 percent of peanuts produced are roasted and salted, either shelled or unshelled. If you've ever eaten peanuts you know how delicious they are. An open bag is an empty bag! You just keep popping these peanuts in your mouth until they're gone.

Peanuts have high food value and are easily digested. They are rich in the B vitamins. You could get very sick if you did not get enough of these vitamins in your diet.

Peanuts are also rich in protein. They have more protein than other good foods like beans, peas, oatmeal, eggs, cheese and fish. We need proteins to build up muscles. There are 2,650 calories in just one pound (.5 kg) of roasted peanuts. That's more than enough energy for a whole day of work, study and play.

Peanuts are also rich in oil. Only the oil of soybeans, sunflower seeds and grape seeds is of better quality. Peanuts are ground up and the oil is squeezed out with large presses. Peanut oil has many uses. The next hamburger you eat may have been fried in it. And the salads you eat may well have dressings made with peanut oil.

The raw, untreated peanut fresh from the plant is not very tasty.

Other Uses. Peanut oil may have been used to make your baseball glove more flexible, or to make your dad's shaving cream or your mother's face powder. And how about that highway you speed over in your family car? Chemicals from peanut oil may have been used to a make the explosives that blasted out the road bed.

Once the oil has been removed from the ground peanuts, a solid cake-like material is left. This cake may be sold as feed for cattle and pigs or it might be used to fertilize farmlands so crops grow better. Perhaps the corn, peas, carrots or other vegetables you might have had for supper last night were grown in soil enriched with peanut cake.

Even peanut shells have value! Pressed into wallboard, they can support the walls of your classroom or home. The plastic cups, bowls, jugs and peanut butter jars in your kitchen are made with the help of chemicals from peanut shells.

Peanut plants beginning to flower.

How to Grow Peanuts

Are you hooked on the wonderful taste of peanuts? Why not grow your own peanut crop in a backyard garden? It would save money and could be a lot of fun.

A peanut farm in Georgia.

The first or primary root of the peanut plant.

Peanuts grow best in the Southern United States where they can get five months of warm, frost-free weather. First, you need the seeds (peanuts). Your parents could buy them from a farm supply store. Or you could order them from a mail-order seed company. A special peanut variety can be bought for planting in the Northern States like Minnesota, Wisconsin, Michigan and New York. In these states, the peanut should first be raised indoors to the seedling stage.

The developement of the primary and the secondary roots and the emergence of leaves.

Fill several large flower pots with rich loose soil. Plant the seeds (peanuts) about two inches (5 cm) deep. Water lightly from time to time. Keep the plants near the window where they get plenty of sunlight. Carefully remove the seedlings when they are about six inches (15 cm) tall and then plant them outside. Ask your parents for space in your family garden. Make sure your peanut garden will get a lot of sunshine. Shade will kill the plants. Also make sure your plot is well-drained. Otherwise the water will collect after a rain storm and drown your peanuts.

The peanut plant leaves as they develop above ground.

Early June is a good planting time. Loosen the soil with a hoe. Then rake it to make it level. Plant the peanuts (or peanut seedlings) about two inches (five cm) deep and ten inches (25 cm) apart. Add some fertilizer so the young peanut plants have nutrients for health and growth.

After about five months, dig up a few peanuts to see if they are ripe. Break open the shells. If the peanuts fill the shells, they are ready to be harvested. Dig up the whole plants. Then wash the dirt off the peanuts with your garden hose.

Stretch out a line between two posts in your backyard. Drape the peanut plants over the line so the sun and wind will dry them. This may take several days.

Once they are dried, pick the peanuts from the plants. Hang the dried peanuts in a mesh bag in the garage, pantry or corner of your kitchen. They will keep well for several

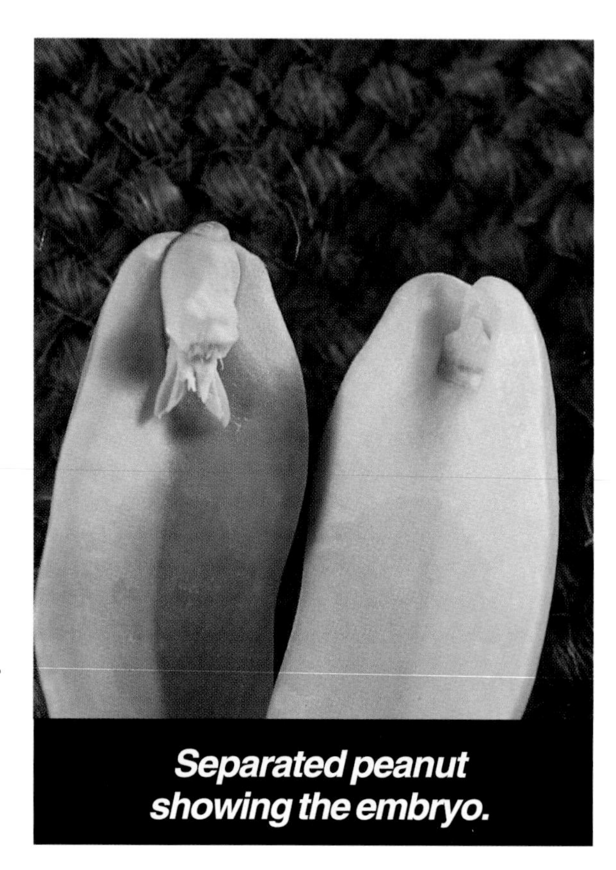

Separated peanut showing the embryo.

months until you want to eat them. To roast the peanuts, ask your parents for a cookie tin. Spread the peanuts on the tin. Place them in the oven at 375 degrees Fahrenheit. Stir the peanuts so they don't burn. They will slowly change color from white to a golden brown. Now it's time to remove them from the oven and have a feast. You'll find that these peanuts which you grew yourself will be the most delicious you've ever eaten!

Fully developed peanuts.

Glossary

Atmosphere the air.

Bacteria microscopic one-celled organisms.

Calories a certain amount of energy found in food.

Carbon dioxide a gas with the formula CO2 which is present in small amounts in the air; it is given off by all living things when they "burn" energy-rich food.

Chemicals basic substances which make up matter.

Chlorophyll green pigment in plants which is needed for photosynthesis.

Chloroplasts wafer-shaped bodies in the plant leaf which contain chlorophyll.

Nectar sweet fluid produced by plants and collected by bees for making honey.

Nitrogen-fixing bacteria found in the root nodules of peanut plants which make it possible for plants to make use of nitrogen in the soil.

Nodules round swellings in the roots of the peanut plant which contain millions of nitrogen-fixing bacteria.

Nutrient a substance which is needed by plants and animals so they can grow and stay healthy.

Ovary the part of the plant flower which makes eggs.

Oxygen a gas needed by all living things so they can use the energy in their foods.

Peanut cake the material which is left after the oil has been squeezed out of peanuts.

Peg the stalk which holds the baby peanut plant and grows down into the soil.

Petals the blade-like parts of a flower; often highly colored.

Photosynthesis the process by which a green plant uses the sun's energy to make food.

Pod the capsule which holds the seeds of a pea, bean or peanut plant.

Pollen a dust-like particle made by the flower which contains a "sperm" cell.

Protein a basic, nitrogen-containing food found in beans, peas, peanuts, meat and other foods which is needed for health and growth.

Sperm a male cell needed for fertilization of the egg.

Stamens club-shaped parts of the flower which make pollen.

Stem the long, slender part of a plant which holds the flower and leaves.

Tap root the large central root of the peanut plant.

Vitamins special substances found in milk, fruits, vegetables and other foods which are necessary for health and growth.

Vitamin B special substances found in peanuts and other foods which are needed to prevent diseases like pellagra and beriberi.

Bibliography

Allen, O.N. and Ethel K. Allen. *The Leguminosae*. Madison, Wis.: The University of Wisconsin Press, 1981.

Askey, Linda C. *Snacks from the Garden*. Southern Living. May 1993, p.72.

Encyclopedia Americana. Entry on Peanuts. Danbury, Conn.: Grolier, 1994.

The Economist. *Peanut Envy*. April 13, 1991, p. 31.

Wilson, Carl L., Walter E. Loomis, and Taylor A. Steeves. *Botany* (Fifth edition). New York: Holt, Rinehart and Winston, 1971.

World Book Encyclopedia. Entry on Peanuts. Chicago: Field Enterprises, 1990.

Index

About the Author

Oliver S. Owen is a Professor Emeritus for the University of Wisconsin at Eau Claire. He is the coauthor of *Natural Resource Conservation: An Ecological Approach* (Macmillan, 1991). Dr. Owen has also authored *Eco-Solutions, Intro to Your Environment* (Abdo & Daughters, 1993), and the Lifewatch series (Abdo & Daughters, 1994). Dr. Owen has a Ph.D. in zoology from Cornell University.

To my grandson, Amati: May you grow up to always appreciate and love nature.
—Grandpa Ollie.